Math Concept Reader

A Nose for News and Numbers

by Linda Bussell

Developed for Harcourt, Inc., by Gareth Stevens, Inc.
This edition published by Harcourt, Inc., by agreement with Gareth Stevens, Inc.

Printed in China

ISBN 13: 978-0-15-360235-1
ISBN 10: 0-15-360235-X

8 9 10 0940 16 15 14 13 12 11 10

Harcourt
SCHOOL PUBLISHERS

Chapter 1: Numbers and Mountain Math

Sarah reads the Sunday paper aloud. She is a reporter for her school paper. Sarah is always looking for story ideas.

She reads about national parks. The parks belong to the people of the United States. They protect natural areas and wild animals. Anyone can visit a national park.

Sarah's mom says that she once visited Yellowstone National Park. Sarah decides to write an article about national parks.

Sarah reads more. Yellowstone was the first national park. President Ulysses S. Grant set aside the land in 1872.

Sequoia and Yosemite in California became national parks in 1890. The General Sherman Tree in Sequoia is the world's largest!

Hawaii's Volcanoes National Park has one of the world's most active volcanoes! It became a park in 1916.

Ohio's Cuyahoga Valley became a national park in 2000.

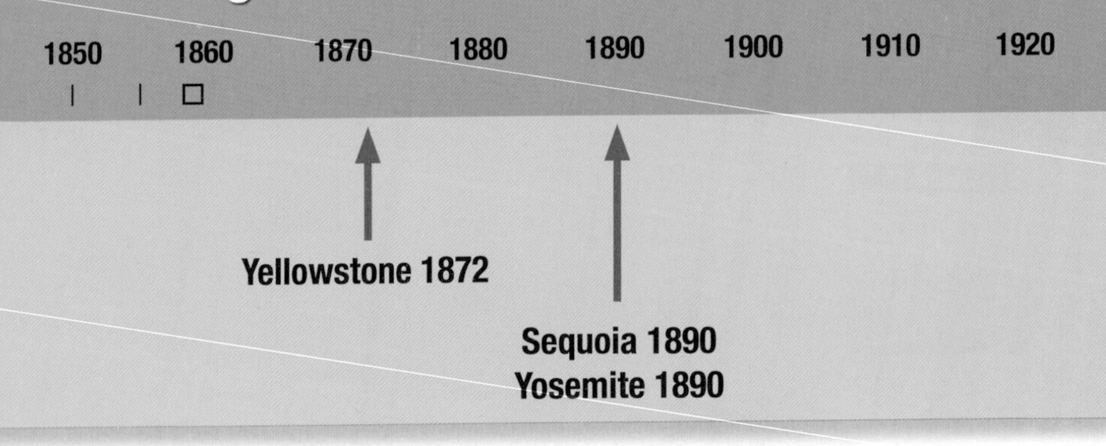

Sarah makes a timeline. A timeline is a number line with dates. She will place dates in order. Her brother Gabe reads aloud.

Mammoth Cave National Park in Kentucky became a park in 1941. Mammoth is the world's largest known cave.

Rocky Mountain National Park was established in 1915. It is in Colorado.

Florida's Biscayne National Park became a park in 1980.

Sarah creates a timeline and marks the years that two national parks were founded.

Sarah draws a line. She puts a mark for every ten years on it. She writes the years above the marks. The line runs from 1850 to 2000.

She places a five-year mark halfway between the ten-year marks.

Sarah enters the dates on her timeline. Where should she put the year 1872? It is between 1870 and 1875. Sarah makes a mark. She labels it "Yellowstone 1872."

She marks Yosemite and Sequoia on the timeline. They are both at 1890.

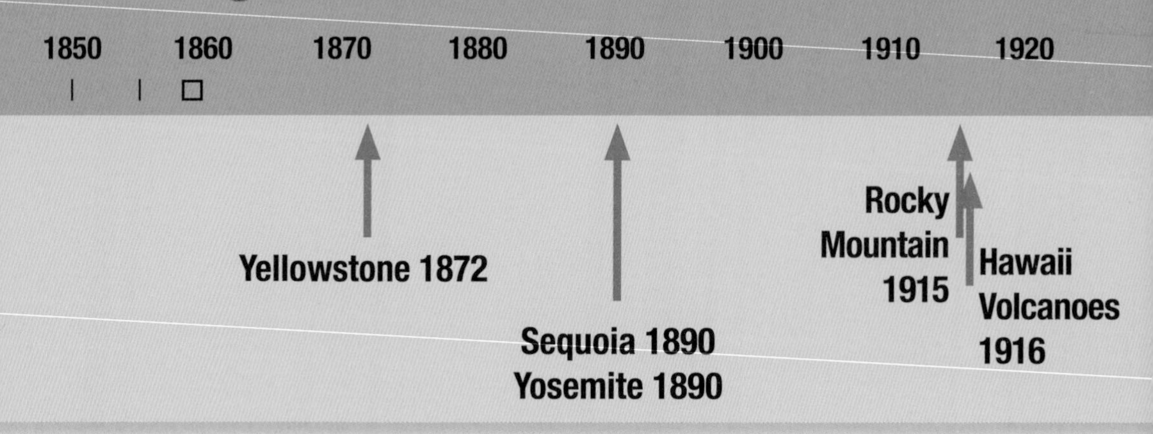

Gabe says that the digits for Biscayne, Yosemite, and Sequoia are the same.

Sarah nods. Then she notices that the place value of the 8 and 9 digits is switched. This makes a difference in the number's value.

Biscayne became a national park in 1980, not 1890. The tens and hundreds digits are reversed. Gabe says 1980 is 90 years after 1890. He writes:

1890 + 90 = 1980

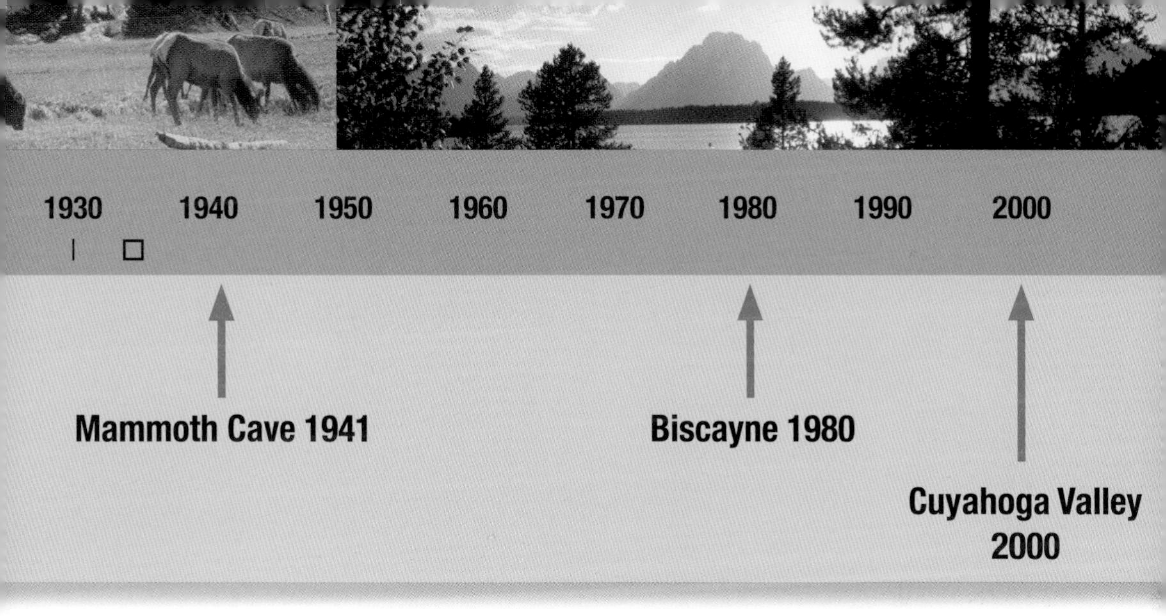

Sarah adds more dates to her timeline, for a total of eight national parks.

Sarah finishes her timeline. Gabe says Sarah's article needs pictures. He helps find them.

Sarah looks at the newspaper article again. The next section is called "Tall Peaks." It is about mountains in the parks. Sarah reads on.

The tallest mountain in North America is Mount McKinley. It is 20,320 feet tall.

Mount McKinley is in Alaska's Denali National Park.

Measured from its base at the ocean floor, Mauna Kea in Hawaii's Volcanoes National Park is the tallest mountain in the world.

Sarah reads about more tall mountains. Mauna Kea is in the Hawaíi Volcanoes Park. It is 13,796 feet tall. Measured from its base on the ocean floor, though, Mauna Kea is the tallest mountain on Earth!

Grand Teton is in Grand Teton National Park. It is 13,770 feet tall.

Mount Whitney is in Sequoia. It is 14,491 feet tall.

Sarah wants to write about the mountains. She draws a table of the mountain heights.

Heights of Mountains in National Parks

Mountain Name	Height
Mount McKinley	20,320 feet
Mount Whitney	14,491 feet
Mauna Kea	13,796 feet
Grand Teton	13,770 feet

Sarah orders the heights of mountains in a table.

Sarah orders the heights of the mountains. She starts with the ten thousands place. Mount McKinley is tallest. It has a 2 in the ten thousands place. The other mountain heights have a 1 in the ten thousands place.

Sarah compares the thousands place. Then she compares the hundreds place. She puts the mountains in order from tallest to shortest. They are Mount McKinley, Mount Whitney, Mauna Kea, and Grand Teton.

Chapter 2: Tall Waterfalls and Counting Creatures

Sarah wants to do more research. She, Gabe and their mom walk to the library. They use computers there to search the Internet. They find books about national parks.

Sarah wants to find other things to compare in the national parks. On the National Park Service Web site, Sarah reads about Yosemite National Park. She learns that Yosemite Falls is the tallest waterfall in North America. Sarah decides to write about the famous waterfalls in Yosemite.

Height of Waterfalls	
Yosemite Falls	2,425 feet
Bridalveil Fall	620 feet
Nevada Fall	594 feet
Ribbon Fall	1,612 feet

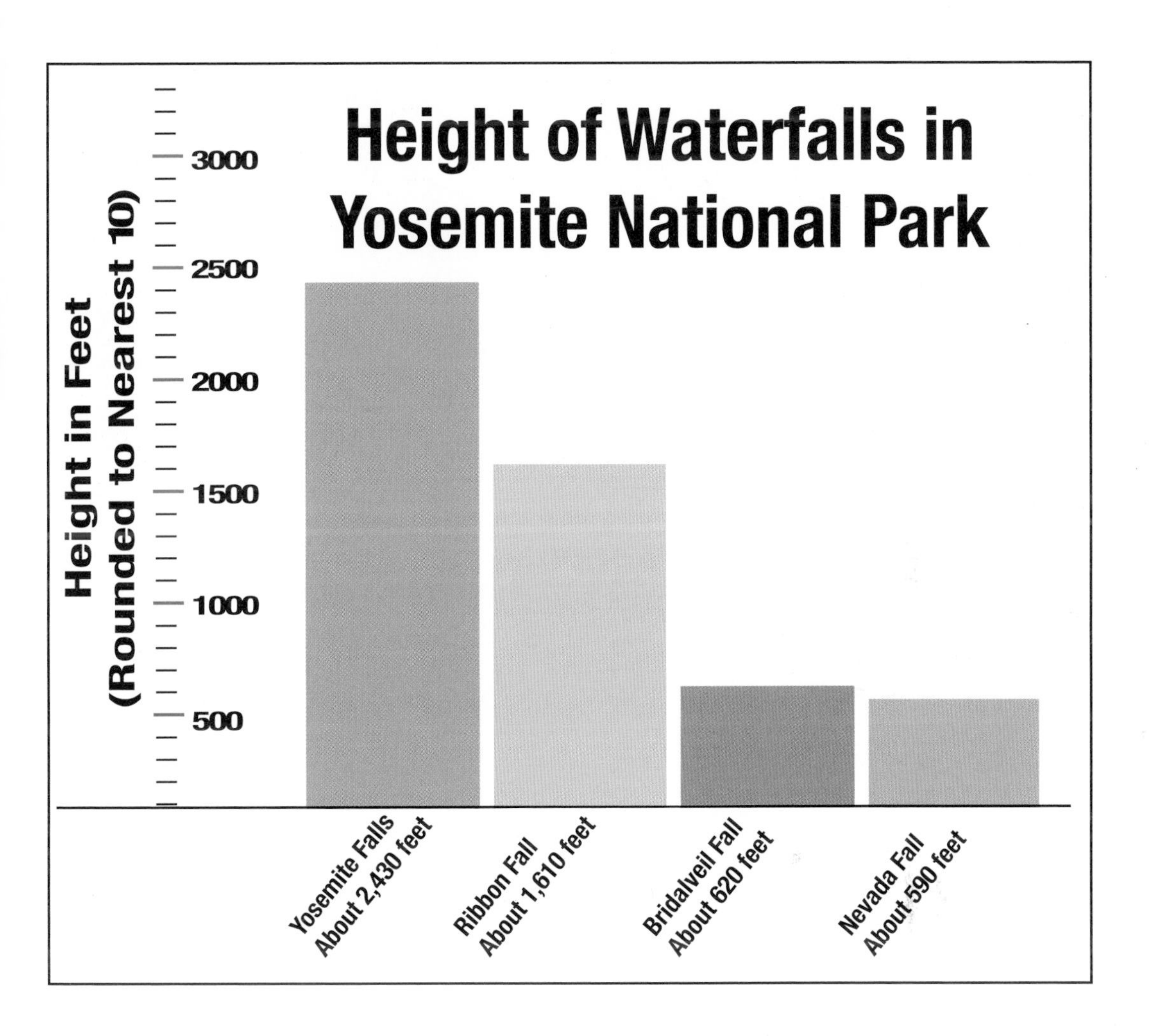

Sarah wants to make a table. It will show the waterfalls in Yosemite in order by height.

"A bar graph would be easier to read," says Gabe. "It is better for comparing the heights. It will be easier to draw if we round the numbers."

Sarah discovers that if she rounds the heights to the hundreds place, it looks like Bridalveil and Nevada Falls are both 600 feet high. If she rounds to the tens place, the differences will show.

National parks are home to many kinds of birds, fish, and mammals.

Finally, they look for information about wildlife. They read that Maine's Acadia National Park has 326 kinds of birds. It has 41 kinds of mammals and 28 kinds of fish. That is 395 kinds of animals in all. They round to the nearest 100. That's about 400 different kinds.

Rocky Mountain National Park has 276 kinds of birds. It has 56 kinds of mammals and 7 kinds of fish. That is 339 kinds of animals in all. They round to the nearest 10. That's about 340 different kinds.

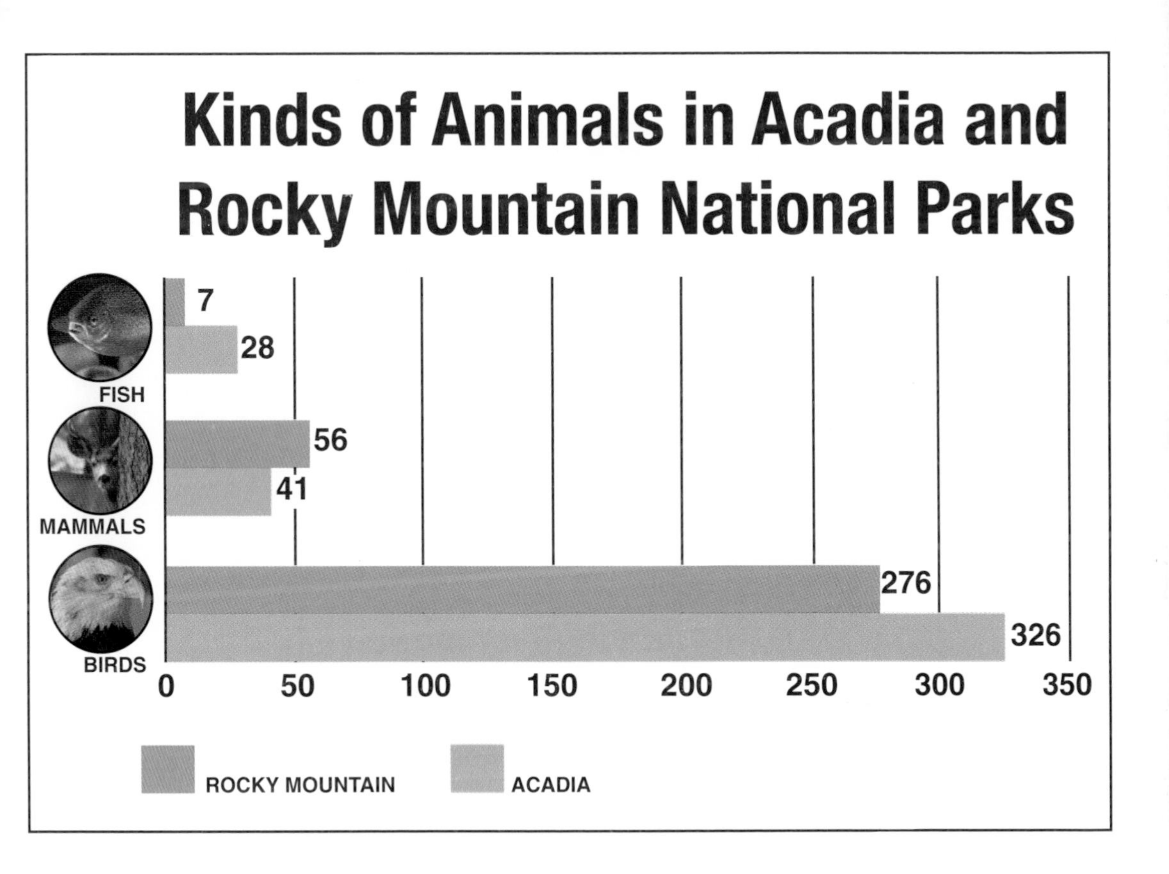

Sarah and Gabe make a bar graph. It shows the numbers of different kinds of birds, mammals, and fish. They can use the graph to compare. It compares the numbers of animals in the two parks.

Sarah makes a bar on the graph for each kind of animal in each park. Gabe wants to show the total number of different animals in each park. Sarah thinks that readers will figure this out. They can use data in the graph.

Chapter 3: Read All About It!

Sarah, Gabe, and their mom return home. They are excited about all they have learned. They get dinner ready. They talk about the article.

Sarah has enough information to begin writing. She says she will call her article "National Parks by the Numbers." Gabe says this is a good title. The article has lots of numbers in it.

Sarah's brother and mother help her finish her article. It is her best one yet!

Sarah organizes her article. She orders and rounds numbers. Ordering and rounding numbers makes it easier for readers to understand the facts.

After dinner, Sarah sketches a picture of what the article will look like. She draws rectangles where the tables, graphs, and photos will go. Gabe draws maps and pictures of animals.

Mom finds a picture of Yellowstone to use. They agree that the article will be Sarah's best yet.

Glossary

order to arrange according to a rule

place value the value of each digit in a number, based on the location of the digit

round to replace a number with another number that tells about how many or how much

timeline a number line with dates

Photo credits: cover, title page, pp. 3 (bottom center), 15 Russell Pickering; pp. 3 (top left, middle left), 4, 5 (both), 6, 7 (both), 9, 13 (middle) National Park Service; p. 3 (bottom left) courtesy of MODIS Rapid Response Project at NASA/GSFC; p. 3 (right) © Kayte M. Deioma/Photo Edit; p. 8 NOAA; pp. 12 (left), 13 (top) U.S. Department of Agriculture; p. 12 (top right, bottom right) National Biological Information Infrastructure.